Series 117

This is a Ladybird Expert book, one of a series of titles for an adult readership. Written by some of the leading lights and outstanding communicators in their fields and published by one of the most trusted and well-loved names in books, the Ladybird Expert series provides clear, accessible and authoritative introductions, informed by expert opinion, to key subjects drawn from science, history and culture.

The publisher would like to thank the following for the illustrative references for this book:
Endpapers: iStockphoto/Duncan1890

Every effort has been made to ensure images are correctly attributed, however if any omission or error has been made please notify the Publisher for correction in future editions.

MICHAEL JOSEPH

UK | USA | Canada | Ireland | Australia
India | New Zealand | South Africa

Michael Joseph is part of the Penguin Random House group of companies whose addresses can be found at global.penguinrandomhouse.com

First published 2019

001

Text copyright © Helen Scales, 2019

All images copyright © Ladybird Books Ltd, 2019

The moral right of the author has been asserted

Printed in Italy by L.E.G.O. S.p.A.

A CIP catalogue record for this book is available from the British Library

ISBN: 978-0-718-18909-9

www.greenpenguin.co.uk

Penguin Random House is committed to a sustainable future for our business, our readers and our planet. This book is made from Forest Stewardship Council® certified paper.

Octopuses

Dr Helen Scales

with illustrations by
Alan Male

Ladybird Books Ltd, London

Eight-armed celebrities

There's something strange about octopuses, something that makes them quite unlike any other animal on Earth. They reach into our imaginations with their suckered arms, grab hold and don't let go.

Sometimes the octopuses we conjure up are charming and companionable. Ringo Starr invited us beneath the waves, to the welcoming seclusion of an octopus's garden. Disney's *Finding Dory* sees the eponymous fish having an adventure with Hank, a grumpy octopus.

Often, though, fictional octopuses are far more frightening. In Jules Verne's *Twenty Thousand Leagues Under the Sea*, a fearsome pack of *poulpes* (French for octopus) attack Captain Nemo's submarine. James Bond tangled with Octopussy's band of female gangsters, all tattooed with the deadly blue-ringed octopus. And author H. P. Lovecraft unleashed Cthulhu, a monstrous, deep-dwelling humanoid with a giant octopus for a head, who has inspired all manner of tentacle-clad characters, from the noodle-faced Ood in *Dr Who* to Davy Jones in *Pirates of the Caribbean*.

Octopus oddities have even led people to believe they hold psychic powers. Paul the Octopus, from a German aquarium, allegedly predicted winners of the 2010 Football World Cup. Given a choice of two food-filled boxes decorated in competing teams' flags, Paul kept picking Germany (perhaps recognizing the bold stripes), and it just happened that Germany kept winning.

Our familiarity with all these fun and formidable octopuses has, at least until recently, been rather at odds with our understanding of the real, living creatures. Gradually we're discovering that fact is often far stranger than fiction.

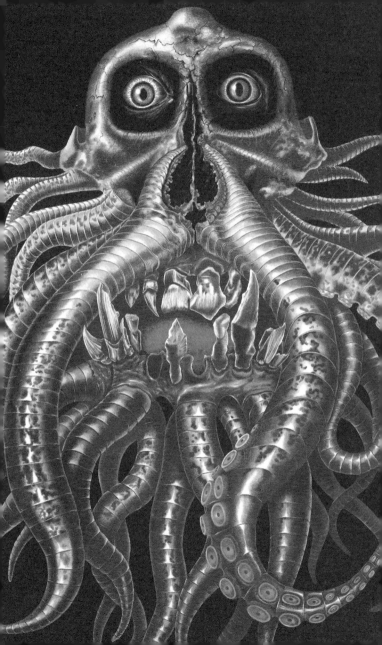

Gods and monsters

Long ago, octopuses began to swim through myths and legends, a symbol of the unseen, mysterious depths. In ancient Fijian mythology, an octopus goddess defeated a mighty shark god who tried to conquer her island. She wrapped him in her powerful arms and as she squeezed he cried for mercy. She released him on condition that he promised to protect her people from shark attacks. And in cold, Scandinavian waters lurks the kraken, inspired in the eighteenth century by octopuses (and perhaps their relative, the giant squid).

The term 'octopus' stems from the past, originating in a Latinized form of the Ancient Greek word *oktopous* (*okto* meaning eight, and *pous* meaning foot). This has led to confusion over which word to use when there's more than one octopus. Some claim it should be 'octopi' (from the Latin), or 'octopodes' (from the Greek). But in fact, it's neither. Whenever a foreign word is adopted into the English language it becomes inflected just like other English words: one octopus, many octopuses.

Ancient stories of octopuses have spilled into artworks and artefacts worldwide. The Minoans of Crete painted jugs and vases with goggle-eyed octopuses 3,500 years ago. Around 400 AD, Moche chiefs in Peru wore sparkling gold headpieces depicting eight-armed deities.

Octopus myths and motifs offer glimpses of the living animals they were originally based on, but these depictions will only take us so far. To truly get to know them, we need to dive into the real world of octopuses.

What is an octopus?

When it comes to spotting an octopus, there's a set of unique features to look for.

First, and most obviously, keep an eye out for lots of limbs. If an animal has eight soft, flexible arms with powerful suckers arranged all the way along them, then it's an octopus. (Squid and cuttlefish have eight arms plus two tentacles, with suckers confined to the tips.) Sometimes an octopus might have fewer than eight arms, if it recently had a run in with a predator: they can bite off an arm to escape capture, then regrow another in its place.

Octopuses use their arms to creep along the seabed, their soft bodies flowing behind. They also swim by jet propulsion, squirting water out of the tube which they breathe water through, pointing it in different directions to steer.

On the inside, octopuses have other unusual characteristics. Peer into an octopus's mouth, keeping your fingers well clear, and you'll see a powerful weapon, its sharp, parrot-like beak, which it uses to chew prey. Octopuses also have a venomous bite.

What's more, they have blue blood and three hearts, one to pump blood around the body and the other two supplying the gills. These cardiac embellishments are probably due to the composition of their blood. Unlike vertebrates, with iron-rich haemoglobin molecules packed into red blood cells, octopuses have a copper-rich equivalent called haemocyanin dissolved directly in their blood, tinting it blue. It's less efficient at transporting oxygen than haemoglobin, so the extra hearts pump more blood, boosting oxygen levels to sustain the octopus's active lifestyle.

Singular softies

One feature all octopuses lack is a backbone. They're invertebrates, part of a vast lineage of cold-blooded animals including jellyfish, crustaceans, insects and spiders.

Within the invertebrates, octopuses are molluscs. Their close living relatives include slugs and snails (the gastropods), clams, oysters and mussels (bivalves), and closest of all are the squid, vampire squid, chambered nautiluses and cuttlefish – all types of cephalopod (pronounced with either a soft or hard 'c', your choice).

Among molluscs and the other invertebrates, octopuses are truly remarkable. They have enormous brains and behave in complex ways that we usually only see in backboned animals, the vertebrates, including us humans.

And yet we last shared a common ancestor with molluscs at least 600 million years ago, back when complex animal life was just getting going. Probably that ancestor was a small, worm-like creature, although we can't be sure: no fossils have been found. Scientists have estimated when they existed by reading 'genetic clocks'. Genetic codes gradually change across generations of all living creatures and by comparing DNA from any two species you can estimate when they shared an ancestor.

Following that ancient split, animals with and without backbones parted ways, forming two distinct groups. Roughly 525 million years ago, there were tiny, eel-like creatures that were probably the first vertebrates. There was also an assortment of swimming, scuttling, shelly invertebrates. This was the beginning of two immense, independent strands of life on Earth that have existed for hundreds of millions of years.

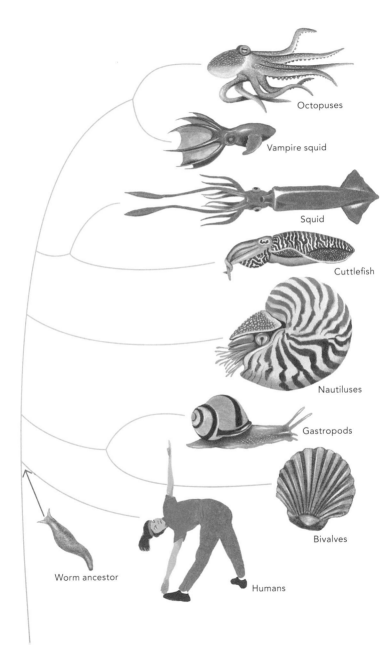

Meet the octopuses

Octopuses first evolved in the sea and there they've stayed and flourished. Across the world's oceans there are roughly 300 known species – none live in freshwater or permanently on dry land – and they come in a huge variety of shapes and sizes. There are whoppers like the giant Pacific and seven-arm octopuses (which in fact have eight arms, but hide one – we'll find out why later). Then there are star-sucker pygmy octopuses, which when fully grown could swim around in an egg cup.

Many species live in warm or temperate waters. They ramble over the seabed in sandy shallows, on rocky and coral reefs, or slink through the shadows of giant kelp forests.

Frigid seas are no barrier to octopuses. Around Antarctica, they hunt for clams, drilling into their shells and injecting venom that works at sub-zero temperatures.

Out in open seas, octopuses spend their whole lives swimming. Blanket octopuses have webs of skin between their arms which flutter like sheets hanging on a washing line. Dumbo and flapjack octopuses swim by beating ear-like flaps (octopuses don't actually have external ears but they can hear, with organs called statocysts). Glass octopuses have transparent bodies and look like they'd shatter if you dropped one.

Most extreme are the octopuses recently spotted near scorching hydrothermal vents, nestled among giant tube worms.

To find out how this diverse crop of modern octopuses evolved, we need to jump back in time to meet some cephalopods that are no longer around.

The age of cephalopods

Early in the history of molluscs the main groups emerged, including bivalves, gastropods and cephalopods. Exactly what the most ancient cephalopods looked like remains contentious. In the Cambrian Period (starting roughly 570 million years ago), there were strange, squid-like creatures called *Nectocaris* that could have been very early cephalopods, but not all experts agree.

By the Ordovician Period, around 492 million years ago, the seas were filling up with shelled cephalopods. There were creatures with enormous straight or gently curved shells. They lay on the seabed, or swam just above it, and preyed on bivalves and trilobites. *Cameroceras* was the biggest, with a ten-metre shell, as long as a London bus. Other species darted around like giant spears and some, with corkscrew-shaped shells, pirouetted up and down. These early cephalopods were the world's first super-predators, long before huge vertebrates evolved.

For hundreds of millions of years, the most abundant and diverse cephalopods were ammonoids. They had coiled shells, some with loops and twists. The biggest was *Parapusozia*, with a shell two metres across, bigger than a tyre on a monster truck. Ammonoids left behind millions of shells that became fossilized, but only in rocks that formed more than 66 million years ago. It was then that ammonoids went extinct, along with the dinosaurs. A giant meteor slammed into the Earth, massive volcanoes erupted, the skies went dark and the seas became caustic and acidic.

But of course, not all cephalopods went extinct. Among the survivors a transformation took place when they gave up a key body part, paving the way for even more dramatic changes to come.

Skinny dipping

At some point, some of the ancient cephalopods lost their shells and went naked. Today, a handful of chambered nautilus species are the only living cephalopods that occupy an external shell. Others now wear their shells on the inside. Cuttlefish have spongy cuttlebones, vestiges of their ancestors' shells, which help them float underwater. Various squid and a few octopuses have remnants of internal shells. But by and large, octopuses go entirely shell-free.

It's tricky to pin down exactly when soft-bodied octopuses first evolved. Being so squashy, they haven't readily fossilized and only a few intact specimens have been found – as unlikely as a sneeze pressed into rock.

Recent studies using genetic clocks have revealed that the modern cephalopod group containing octopuses, squid and cuttlefish – collectively known as the coleoids – probably first arose around 160 to 100 million years ago.

This was during the so-called Mesozoic Marine Revolution, a time when the oceans were becoming a distinctly dangerous place. Giant marine reptiles like ichthyosaurs, placodonts and mosasaurs were prowling the seas, as were ferocious sharks.

Meanwhile prey species were busy evolving means of avoiding them. Some invertebrates responded by building thicker, heavier shells, but the cephalopods did the opposite. They lightened their loads, reducing and losing their shells, giving them greater agility and speed to escape the hunters.

Opting for a shell-free life also opened up an ocean of other possibilities for the cephalopods.

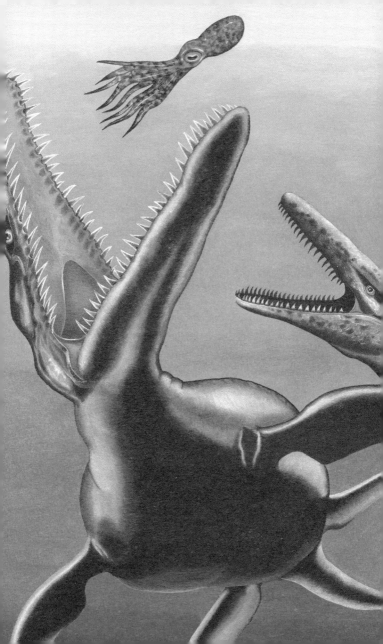

Bendy bodies, big brains

Losing their shells allowed cephalopods to become supple shape-shifters. With rubbery bodies and no hard parts except for their beaks, octopuses can take on almost any imaginable shape. They can pour themselves into empty beer bottles and escape from the decks of fishing boats, squeezing through impossibly narrow holes.

One theory suggests that losing their shells was an important evolutionary step towards the octopuses' large brains. It takes quite some doing to control eight gangly arms. A sophisticated nervous system is vital for muscle coordination and to make the most of this pliable anatomy.

Octopuses have the largest brain relative to body size of any invertebrate, bigger even than some vertebrates. They evolved from ancestors with multiple ganglia, tight knots of neurones, distributed around their bodies. In modern octopuses, enlarged ganglia form a lobed brain which, oddly enough, surrounds the oesophagus. When an octopus swallows, food passes right through its brain.

But octopus brains tell only part of the story. In humans, most of our 100 billion neurones are clustered inside our brains. More than half of an octopus's 500 million neurones are located in its arms. These semi-autonomous limbs can, to some extent, think for themselves.

Often octopuses will reach out and touch a scuba diver. But don't assume they adorably want to hold hands – they're probably just deciding if the diver is worth eating. Each sucker contains 10,000 neurones, letting it taste and touch. Even a dismembered octopus arm carries on grabbing things for hours, passing them towards where the mouth used to be.

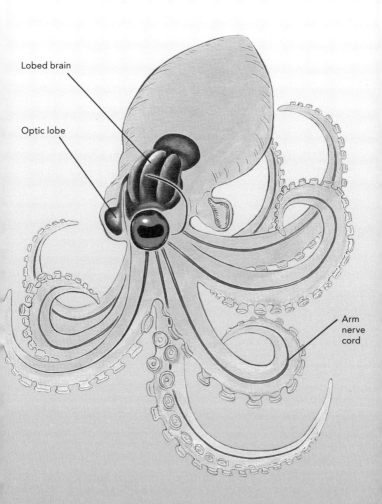

How the octopus got its smarts

Once octopuses developed an advanced nervous system to coordinate their soft bodies, this may have primed them to co-opt and develop their brains to do other things – to remember, to solve problems, to navigate and to become fast-thinking hunters.

Many of these abilities were discovered by chance by scientists who spotted octopuses doing surprising things.

In 2009, Australian biologist Julian Finn came across an octopus in Indonesia carrying two halves of an empty coconut. It strutted across the seabed, using some of its arms as legs, before stopping and sealing itself inside the two coconut halves. Other octopuses have since been spotted doing the same trick, sometimes with empty clam shells.

These bizarre performances demonstrate several important things about the coconut-carrying octopuses. Assembling and then disassembling a mobile shelter from separate items is an example of complex tool use – something it was thought only some mammals and birds can do. Even more remarkably, the octopuses have a sense of foresight: they know the two shells might come in handy at some point in the future. They have a plan.

These and many other clever behaviours are the upshot of an entirely separate evolutionary experiment in intelligent life, far removed from our conventional idea of smart vertebrates (including ourselves). Octopuses are so smart, in fact, that they are designated 'honorary vertebrates', and scientists need a licence to work with them in captivity. Such studies are revealing further facets of the octopuses' outstanding, invertebrate intelligence.

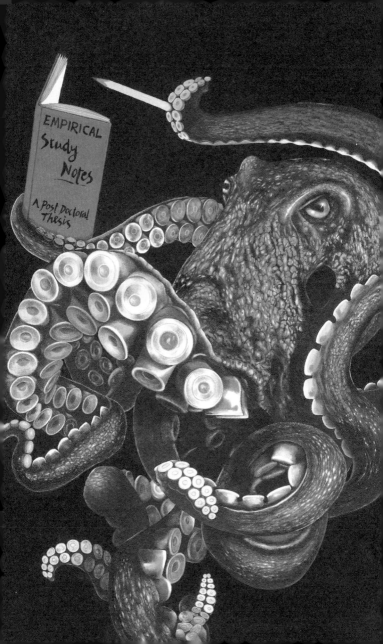

Tank tricks

Captive octopuses are notorious for escaping and going on foraging raids into nearby tanks, feasting on other animals before returning to their home tank.

And captive octopuses show more obvious signs of unusual intelligence, quickly figuring out what to do in unfamiliar circumstances. There's the case of at least two octopuses, in different places, repeatedly and deliberately squirting water at a bright light bulb until it went out. They both seemed to understand what they were doing.

Octopuses are also known to take against certain people, intentionally drenching them with a squirt of water whenever they walk past. Studies show octopuses can recognize human faces. This could be linked to the fact that they also seem to recognize each other and remember octopuses they've previously hung out with.

They can also learn to unscrew containers to get at food inside (some even screw them back up afterwards); they can take apart Lego sets and within an hour work out the push-and-twist knack of opening a childproof medicine bottle.

It's possible to train an octopus. When offered a reward of food, they can learn to pick out different targets either by sight or touch, and they learn to navigate through mazes. Not all comply though; some stubbornly refuse to play along, even snapping off the lever that provides food.

All of this could be explained by the octopuses' curious, exploratory nature, which fits with their hunting behaviours in the wild. Sometimes, though, octopuses do things that leave scientists baffled.

Personalities and play

Several decades ago, octopus researcher Jean Boal encountered an octopus that refused to eat its dinner. Fresh crab is most captive octopuses' meal of choice, but for convenience they're often weaned on to frozen shrimp. One day at feeding time, one of Boal's octopuses conspicuously held up its frozen titbit, looked her right in the eye and shoved the shrimp down the outflow pipe in its tank. The only explanation Boal has is that the infuriated octopus was showing it would much rather eat something else.

Aquarium keepers often give octopuses names reflecting their characters – some are angry, some are shy, bold or cheeky. Variations in personality could allow them to learn and adapt quickly to new situations.

A further sign of octopus intelligence is the fact that some of them like to play. Such spontaneous activities serve no obvious purpose other than pure enjoyment. It's hard to imagine a slug or snail doing something just for fun, and yet that's precisely what some of their close relatives seem to do.

A classic example comes from Canadian octopus biologist Jennifer Mather and Roland Anderson, keeper at the Seattle Aquarium, who presented octopuses with a floating pill bottle. At first, each octopus put the bottle in its mouth (presumably checking whether it was edible). Then several of them squirted water jets at the bottle, pushing it across the tank, and waited as it bobbed back towards them in the current from the aquarium's intake valve. Two octopuses spent several minutes blowing the bottle across the tank, over and over, like bouncing a ball against a wall.

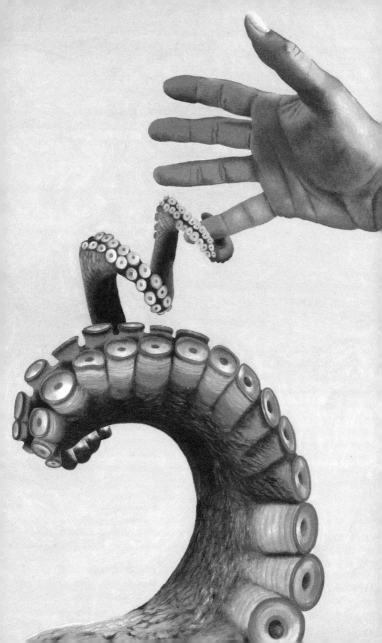

The puzzle of the argonauts

Octopuses lost their shells millions of years ago but one small group returned to a hard life. For centuries, scientists have pondered over tiny octopuses called argonauts that swim through open seas inside delicate shells. Some suggested that argonauts attack and eat other shelled creatures, then float off across the waves, using the empty shell as a boat and hoisting two flattened arms as sails.

Eventually it was pioneering scientist Jeanne Villepreux-Power who uncovered the truth about argonauts. In the 1830s, on the island of Sicily, she invented the aquarium tank and conducted ingenious experiments which revealed that argonauts aren't pirates but make shells themselves, using the ends of two arms.

Villepreux-Power saw female argonauts using their shells as mobile brood chambers, to carry and rear their young. But she saw no sign of the males.

We now know that male argonauts are minute compared to females, and they don't have shells. They swim around, hoping to come close enough to a female to donate a detachable arm that's laden with sperm (this had been previously misidentified as a parasitic worm). She then carries around his appendage, and several from other males, using them as required.

Biologist Julian Finn has discovered another use for the argonauts' shells. Recently, he released several live female argonauts underwater and watched as each one swam upwards, popped her shell above the waterline and trapped an air bubble inside, before jetting off into the distance. Like their cousins, the chambered nautiluses with gas-filled shells, argonauts use their shells as energy-saving buoyancy devices that stop them from sinking.

The hunters become the hunted

For most octopuses, life without a shell has its benefits but also a major drawback: they can easily become someone else's dinner.

Hordes of aquatic predators like to snack on these soft cephalopods. Dolphins have learned how to fling and kill octopuses before eating them, to avoid being choked by their clinging suckers.

Octopuses have various tactics that make up for their lack of bodily protection. They are venomous, most notoriously the blue-ringed octopuses. Bacteria in their salivary glands make tetrodotoxin, the same toxin that makes pufferfish deadly to eat. There's enough in a single blue-ringed octopus to kill ten people.

Meanwhile, blanket octopuses tear off stinging tentacles from Portuguese men-of-war and use them as weapons.

Octopuses also squirt ink, a mixture of mucous and the dark pigment melanin. The ink cloud may confuse predators by creating an octopus-shaped shadow while the real one escapes. It could also taste bad to fish.

Another part of an octopus's survival kit is its intelligence. Emerging from their dens to forage, octopuses keep track of their meandering path – perhaps memorizing local landmarks – and dash straight back to safety if a predator appears. A recent BBC film shows a shark catching an octopus, then moments later spitting it out because the crafty cephalopod shoves several arms into the shark's gills and it can't breathe. And in a curious throwback to their ancestors, octopuses sometimes hold empty sea shells in their suckers forming a full shell suit that provides temporary protection and hides them on the seabed.

Disappearing acts

A decade ago, veteran octopus researcher Roger Hanlon was diving in the Caribbean when something made him scream. He caught the moment on camera: a seaweed-covered rock suddenly reveals its true identity – a large octopus. It blanches white, then takes off in a cloud of ink. It was the fastest, most exquisite example Hanlon had ever seen, and filmed, of an octopus using its colours and texture to hide, then escape.

Mesmerizing displays scintillate across octopuses' bodies, the most complex camouflage of any animals. It's another strategy these soft creatures use to avoid being eaten.

Second by second they change their skin's appearance, an impressive feat made possible by their sophisticated nervous systems. They constantly watch their surroundings and decide which body pattern will work best.

An octopus's kaleidoscopic colours are produced in layers of skin cells called chromatophores, which are star-shaped and filled with coloured pigment granules. Nerves from the brain directly control muscles that shrink or stretch these skin cells, hiding or revealing colours. Further skin layers combine to produce iridescent colours, with tiny mirror-like particles that reflect and refract light. Muscles also draw the skin surface into lumps and bumps, called papillae, that mimic rough pebbles or tufty seaweeds.

Octopuses often change shape and colour, pretending to be other things, like a rolling stone or a clump of drifting seaweed. The true masters of disguise are the mimic octopuses. They dress up like a host of other animals to confuse predators and pretend they're more dangerous than they really are. One moment they're a sea snake, next they're a venomous lionfish.

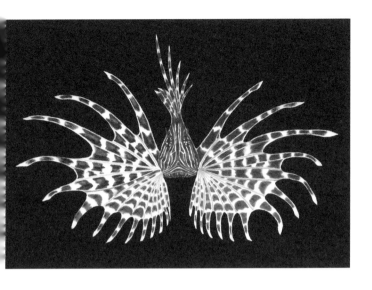

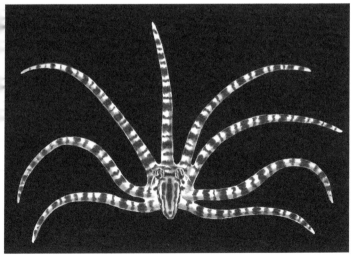

Look into my eyes

Bizarrely, despite their vivid body patterns, it's long been thought octopuses are colour-blind. Octopus eyes are similar to our own; they have a lens, iris and a retina packed with light-sensitive photoreceptors, but there are important differences (highlighting the fact that they evolved separately). Octopuses don't have a blind spot like people do, because their optic nerve is wired up behind the retina, while ours passes in front, like a camera strap falling in front of the lens.

Octopuses have only a single type of photoreceptor. Humans usually have three. Each one responds to certain wavelengths of light (roughly speaking, blue, green and red) and together they generate nerve signals which our brains interpret as colour images. It's clear octopuses haven't evolved the same colour vision as ours, and in lab studies they show little sign of being able to tell apart different colours.

But octopuses have another trick up their sleeves. Roger Hanlon and other cephalopod researchers recently discovered that octopuses may see with their skin. They have photoreceptors spread across their bodies (although still only one type) which may detect light sweeping over them. This information could help them adjust their camouflage.

We also shouldn't entirely dismiss the possibility of octopus colour vision. It could be that their colour-making skin cells, the chromatophores, act like filters, altering the light that falls on photoreceptors and allowing them to respond to particular colours.

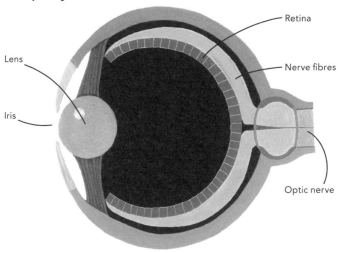

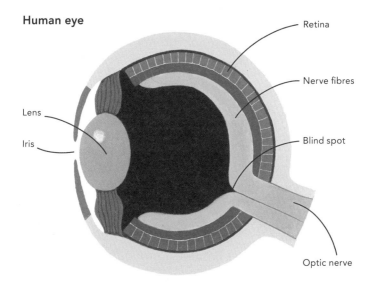

Colourful conversations

Cephalopods use patterns to send messages to other animals, as well as for camouflage.

When Roger Hanlon's Caribbean octopus suddenly turned white, it was saying, 'I've seen you, and I'm large and fierce!' This common display, often combined with splayed arms, startles a predator, giving the cephalopod a chance to escape. Blue-ringed octopuses usually stay hidden on the seabed, but when disturbed they flash sixty shiny blue rings, saying, 'Back off, I'm deadly!'

Cephalopods also employ patterns when in attack mode. Dark waves flickering across a cuttlefish's body seem to mesmerize their prey. Glowing sucker octopods, unsurprisingly, have bioluminescent suckers, which may lure in prey in the dark depths.

Being largely silent, cephalopods also use patterns to talk to each other. Male Caribbean reef squid turn red to attract females and white to fend off males. A pack of Humboldt squid flash red and white all-over, perhaps coordinating their hunt. Oval squid have a repertoire of twenty-seven skin patterns, which could be a simple form of language.

Given the direct connection between the octopuses' colour-making chromatophores and their nervous systems, it's possible their patterns could reflect more subtle emotions, but for now this remains another cephalopod mystery.

Until recently, it was assumed that octopuses are solitary, with little need for cultivated communication. Put two together in an aquarium and one tends to eat the other. However, as more people spend time watching octopuses it's becoming clear that they sometimes learn to get along.

Discovering Octopolis

In 2009, scuba-diver Matthew Lawrence saw something highly unusual in Jervis Bay, near Sydney, Australia. Roaming around a swathe of empty scallop shells were at least a dozen octopuses. Some were wrestling. Some sat still and reached out to touch other octopuses passing by.

Since then, Lawrence has returned many times with a team of scientists, including David Scheel and Peter Godfrey-Smith, to study the site now known as Octopolis. Hours of diving and film footage from cameras left on the seabed have revealed details of this rare octopus aggregation, including their use of colour and posture to communicate. Aggressive males turn dark and loom over others. If their rivals don't become pale, it's likely a fight will break out.

A similar octopus gathering was found nearby in 2016 and named Octlantis. Here, on one occasion, a male yanked another from its den; it fled and found another hiding place, only to be evicted again by the same assailant.

Octopolis may have begun when an unidentified 30cm object, possibly something metal, landed on a scallop bed. This offered shelter for a few octopuses, which feasted on scallops and piled up empty shells. The shell debris then provided perfect material for other octopuses to build dens. By chance, the site combined two octopus necessities: plenty of food and somewhere to hide.

Studies continue at Octopolis and Octlantis and mysteries still abound. When they reach out and give high-fives, do octopuses recognize each other? When they excavate their dens and toss out shells, are they deliberately trying to hit another octopus?

Making more octopuses

Whether they live alone or in quarrelsome 'gardens', there comes a time in every octopus's life when they have to find a mate. For many species, copulation is a rather remote affair. A male keeps his distance, probably to avoid getting eaten by the female, and reaches out with a single, sperm-bearing arm. (Fighting males often try to rip off each other's mating arms and some, including the seven-arm octopus, tuck their important appendage away and only unfurl it during mating.)

Octopus sex is sometimes more intimate. In the 1970s, reports emerged from Panama of octopuses mating beak to beak. Then in 2012, rare live specimens were found of this elusive species, the larger Pacific striped octopus.

Watching these octopuses in tanks, researchers have seen male-female pairs occasionally sitting in the same den and sharing food, nibbling on a shrimp with their arms aligned, sucker to sucker.

Most dramatic is when the octopuses mate in a tangle of arms. The female, usually the bigger of the pair, often wraps the male up in the web of skin stretched between her arms. Sometimes the male releases a puff of ink (no one knows why).

Most octopuses have a single shot at passing on their genes to the next generation. After they've mated, males stop eating and a few days or weeks later they die. Females hang around for longer, laying and tending their eggs, usually inside a den on the seabed (although glass octopuses brood young inside their bodies, and argonauts inside their shells). As soon as the young hatch and swim off, their mothers die.

Live fast, die young

It's a peculiar fact that octopuses live such splendid lives, investing energy in their big brains and complex behaviours but then die after only a year or two. One possible explanation focuses on the octopuses' shell-free existence.

Genetic mutations that happen to act late in life don't normally have any effect because by that stage most animals carrying the mutated genes have already died from some other, outside cause – most likely from being eaten.

Imagine an octopus gene mutates so that, after some time, it makes them go blind. Most octopuses will have been caught by a predator long before their eyesight falters, so the mutated gene isn't weeded out of the population. Instead it's passed on to the next generation and quietly spreads.

There are also 'buy now, pay later' mutations, which are useful early in life (say, by speeding up an octopus's colour changes) but have a cost later on (perhaps the breakdown of chromatophores). These mutated genes accumulate even faster in a population.

Despite their camouflage and intelligence, soft-bodied octopuses' lives are still fraught with danger, and mutations like these may have piled up, making it appear that octopuses are preprogrammed to age, and drastically shortening their lifespans. If an octopus is lucky enough to avoid getting snagged by a shark or flung by a dolphin, it will soon die anyway from these late-acting mutations.

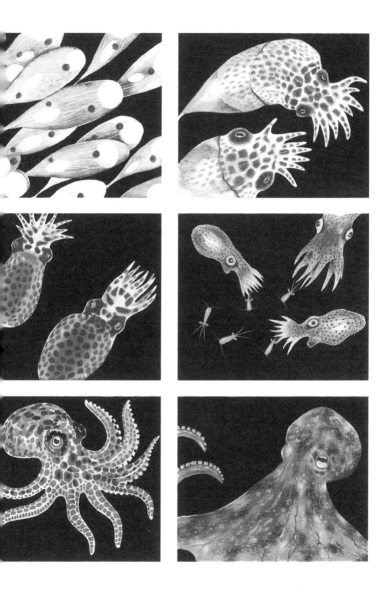

Breaking the rules

Not all octopuses have short lives, though. Things can be different down in the cold, dark depths. Recently, a Remotely Operated Vehicle (ROV) beamed up images from 1,397 metres down off the Californian coast of a female octopus guarding her clutch of eggs, clinging to an underwater canyon. The science team sent the ROV back eighteen times to the same spot and each time found the octopus still watching over her eggs (distinctive scars showed this was the same individual). She stayed there for four and a half years.

This is the longest ever period of egg-brooding reported for any animal. It takes this long for the eggs to hatch probably because the cold water (between 2 and 3°C) slows their development. The chill also slows the mother's metabolic rate, helping her survive for years, perhaps without any food. The Californian octopus showed no signs of eating and ignored pieces of crab offered to her on the ROV's robotic arm.

In shallow seas, female octopuses usually brood eggs for roughly a quarter of their lives. If the same applies to this deep-sea octopus, then she could have lived into her teens.

Deeper-dwelling octopuses may avoid the effects of ageing because their realm is generally safer, with fewer predators than surface seas. Genetic mutations that accelerate ageing may not build up in these species.

This raises a tantalizing possibility. What if other octopus species somehow evolve tactics to outwit predators, at least more effectively than they do now? Could they also push back the signs of ageing? Imagine what octopuses might get up to, with their big brains and curious natures, if they were able to spend longer roaming the oceans.

Proliferating cephalopods

Even though most octopuses don't live long, as a group they're doing very well. Available data show that over the last sixty years octopuses and other cephalopods have increased in abundance worldwide.

Cephalopods are rapidly adapting to our changing world, helped perhaps by their swift life cycles. Warming seas, driven by climate change, could cause them to grow and reproduce even faster, giving them an edge over their competitors. Changes in habitats and ocean currents, and an increase in extreme weather events could shift the balance in favour of cephalopods.

The loss of other species may also play a part. When fish populations crash from overfishing, cephalopod numbers often shoot up. Around Elephant Island near Antarctica, icefish numbers collapsed in the 1980s. Since then, icefish haven't returned but octopus numbers are much higher than in nearby areas. It's possible octopuses are doing well because they're no longer being hunted by icefish.

Time will tell how octopuses will fare in the future. It's likely that fisheries will increasingly target abundant cephalopods, so they too could spiral into decline.

Some octopuses seem safe from human activities – though perhaps only for now. In 2016, scientists found ghostly white octopuses beyond the reach of fisheries, 4,000 metres down in the Pacific. Dubbed 'Casper', these octopuses lay eggs on sponges growing on potato-sized rocks that take millions of years to form. These rocks are rich in precious metals and may soon be dragged up from the deep to make electric-car batteries, solar panels and smartphones. If deep-sea mining goes ahead, Casper could lose her nursery bed.

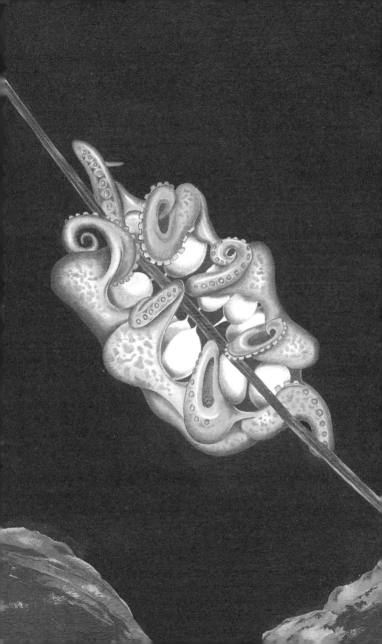

My pet octobot

There are many ideas we can borrow from octopuses and apply to the human world. Engineers are seeking octopus inspiration to spark a new generation of robots. Forget boxy, metal machines and imagine instead a soft, squashy robot that crawls, swims and carefully picks things up.

Prototype octobots are being built that mimic various aspects of the octopuses' strange anatomy. Bionic tentacles have suction cups that gently wrap around objects. Octopus biologist Roger Hanlon recently teamed up with engineers to make a programmable material inspired by octopus skin: silicone rubber sheets inflate into 3D shapes, then flatten again on command. It could lead to 'exosuits' for soldiers or vehicles that adjust their appearance to match the scenery at the flick of a switch.

Professor of biorobotics Cecilia Laschi brought live octopuses into her lab on the Tuscan coast in Italy to study how they move. Her team built robots that crawl in a similar way, stretching and shortening their arms and pushing off the sea floor. Swimming robots have arms preprogrammed to respond individually with no additional computing power from a central machine, similar to the semi-autonomous arms of a real octopus.

Soft robots should work better in unpredictable environments compared to traditional machines that easily get stuck behind unexpected obstacles. Octobots could mould themselves to their surroundings. Maybe they'll assist with delicate surgery, go on rescue missions to dangerous places or live in our homes doing our chores. And who knows, maybe one day someone will make an octobot that's even more like a living octopus and can think for itself.

Being an octopus

What's it like to be an octopus? Are they conscious, self-aware beings? For now, we simply don't know.

Consciousness is a slippery concept – that feeling of being immersed in your own inner world, with a stream of thoughts making you undeniably *you*. It's hard enough to define and measure consciousness in humans, let alone in other animals that we can't talk to. But when you spend time with an octopus, it's hard to shake off the sense that it's looking back with knowing eyes.

Certainly, we can't rule out the possibility of octopus consciousness. In 2012, a group of prominent neuroscientists signed the Cambridge Declaration on Consciousness, stating that humans are not unique in having the nerves and brains needed to generate conscious feelings. All mammals and birds are similarly well endowed, as are octopuses.

If octopuses are self-aware, their consciousness probably evolved in a very different way to our own. With their dispersed nervous systems, octopus consciousness could well manifest itself in a way that we humans wouldn't recognize. It's even possible that there is such a thing as being not just an octopus but an octopus's arm.

And if octopuses are conscious beings, what should we do about it? How would it change our attitudes towards them? If it turns out octopuses feel joy and anger, contentment and anxiety, should we continue to keep them in captivity? Should we still eat them? These are yet more questions that have no definite answers, but surely they're worth thinking about.

Aliens on Earth

The longer we watch octopuses, and the more we learn about them, the more obvious it becomes that they're the closest to intelligent aliens we can meet, right here on Earth.

With their shape-shifting bodies, skin that can see and arms with minds of their own, octopuses teach us about the extraordinary possibilities of life.

Gazing across a 600-million-year divide, we're stripped of any notion that ours is the only way to be brainy. Humans and our backboned relatives are not the only ones with nervous systems capable of intelligent thought. Here is living proof that complex minds have arisen at least twice, once among the vertebrates and then again, entirely separately, among the octopuses. And if it's happened here twice already, who's to say intelligent life couldn't evolve many more times, elsewhere in the universe?

Scientists are hunting for clues to the origins of intelligence. Octopuses can help reveal which parts of nervous systems are fundamental to intelligence and which are interchangeable.

We're still a long way from fully comprehending the lives of octopuses. Only in 2017 it came to light that they edit large numbers of genes in their nerves, on a scale far greater than any other species. Why they do this, and what the consequences are, we don't yet know. But their ability to switch around their DNA code could explain how octopuses adapt quickly to the changing environment and, perhaps, how they came to be Earth's smartest invertebrates.

No doubt the octopuses have many more surprises in store for us, as we tumble ever deeper into their world.

Further reading

Peter Godfrey-Smith *Other Minds: The Octopus, the Sea, and the Deep Origins of Consciousness* (William Collins, 2017)

Sy Montgomery *The Soul of an Octopus: A Surprising Exploration into the Wonder of Consciousness* (Simon & Schuster, 2015)

Roger T. Hanlon, John B. Messenger *Cephalopod Behaviour* (Cambridge University Press, 1998)

Roger T. Hanlon, Mike Vecchione, Louise Allcock *Octopus, Squid and Cuttlefish: A Visual, Scientific Guide to the Oceans' Most Advanced Invertebrates* (University of Chicago Press, 2018)

Danna Staaf *Squid Empire: The Rise and Fall of the Cephalopods* (The University Press of New England, 2017)